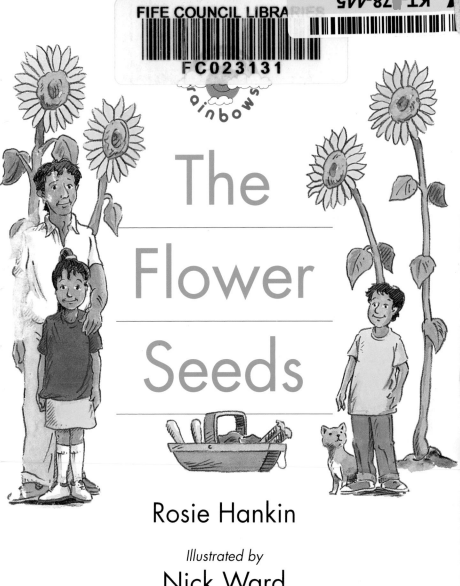

The Flower Seeds

Rosie Hankin

Illustrated by

Nick Ward

Dad is going to plant seeds in the garden.

4

5

These seeds will grow into
the biggest plants in the
garden.

Digging lets more air into the soil and this helps the seeds grow.

9

Now the seeds can be planted.

The seeds will grow roots in
the soil.

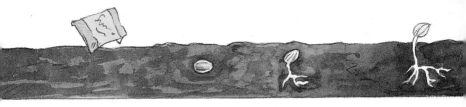

The seeds have sprouted.
Now they are called
seedlings.

13

The seedlings are growing quickly.

14

15

17

The young plants need support.

I'm tying the stems to these stakes.

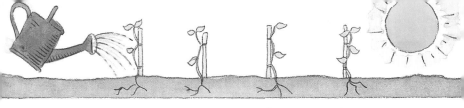

Plants need food from the
soil. They also need sunlight
and water.

Watch out!

19

Soon you will see what kind of plants you have grown.

They're sunflowers.

The flower heads are as big as dinner plates.

The sunflowers are only twenty weeks old but they are very tall.

They are taller than Dad.

24

25

26

The bees are collecting a
sweet liquid called nectar.

The sunflowers have lost all their petals. The flower heads have dried.

Look, they're full of seeds.

28

We'll plant the seeds
next year.

29

Look at this picture of the stages of a sunflower. Can you point to the seedling, roots, stem, leaves, bud, flower head, seeds and petals?

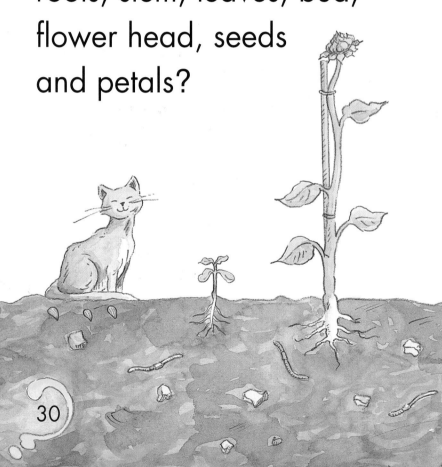